OVERBURDEN CONVEYOR BRIDGE F60

BY QUINN M. ARNOLD

CREATIVE EDUCATION • CREATIVE PAPERBACKS

Published by Creative Education and Creative Paperbacks
P.O. Box 227, Mankato, Minnesota 56002
Creative Education and Creative Paperbacks are imprints
of The Creative Company
www.thecreativecompany.us

Design by The Design Lab
Production by Chelsey Luther
Art direction by Rita Marshall
Printed in the United States of America

Photographs by Alamy (Agencja Fotograficzna Caro, zerega),
Corbis (PATRICK PLEUL/epa), Creative Commons Wikimedia
(Anagoria, A.Savin, Onkel Holz, Michael F. Mehnert), Getty
Images (delectus, PATRICK PLEUL/Stringer)

Library of Congress Cataloging-in-Publication Data
Arnold, Quinn M.
Overburden conveyor bridge F60 / by Quinn M. Arnold.
p. cm. — (Now that's big!)
Includes bibliographical references and index.
Summary: A high-interest introduction to the size, speed, and
purpose of one of the world's largest mobile machines, includ-
ing a brief history and what the future holds for the Overburden
Conveyor Bridge F60.

ISBN 978-1-60818-716-4 (hardcover)
ISBN 978-1-62832-312-2 (pbk)
ISBN 978-1-56660-752-0 (eBook)
1. Earthmoving machinery—Juvenile literature. 2. Coal-mining
machinery—Juvenile literature.

TA725.A74 2016
621.8/65—dc23   2015045211

CCSS: RI.1.1, 2, 3, 4, 5, 6, 7; RI.2.1, 2, 4, 5, 6, 7, 10; RF.1.1, 3, 4;
RF.2.3, 4

First Edition HC 9 8 7 6 5 4 3 2 1
First Edition PBK 9 8 7 6 5 4 3 2 1

TABLE OF CONTENTS

What machine is more than a quarter mile (0.4 km) long? The Overburden Conveyor Bridge F60 is the world's biggest movable machine. It digs up overburden in open-pit mines.

The German company TAKRAF built F60s out of steel.

An F60 weighs 14,990 tons (13,600 t). It is 787 feet (240 m) wide and 262 feet (80 m) tall. It is powered by electricity.

262 ft (80 m)

The Statue of Liberty

This F60 worked in the coal mines of Lusatia, Germany.

Five F60s were built
from 1972 to 1991.
They are used in
German coal mines.
F60s dig up dirt and
rocks to reach coal.
They dig down almost
197 feet (60 m).

9

When it is digging, an F60 moves slower than a sloth.

F60s move on railroad tracks. They have 760 wheels! But they are very slow. The fastest speed an F60 can go is half a mile (0.8 km) per hour.

The F60 has been nicknamed the "Lying Eiffel Tower of Lusatia."

The last F60 to be built was used for only 13 months. Then that mine was shut down. In 2002, it opened to visitors. Thousands of people tour the F60 each year. They go to events in the old mine, too.

The conveyor belts move along at 22.4 miles (36 km) per hour.

Four F60s are still used in coal mines today. It takes 25 people to run each machine. One end of an F60 scoops up overburden. Nine conveyor belts move the load to the other end. It is dumped onto the ground below.

Most of Germany's coal is turned into electrical power.

An F60 can dig up 17,057 cubic feet (483 cu m) per minute. It could fill 11 Olympic-sized swimming pools in an hour!

Many big machines work with the F60 in German coal mines.

Germany plans to use less coal in the future. No one knows what will happen to the monster machines if the mines close. For now, the F60s keep on digging.

BLUE WHALE
←···· *100 ft (30.5 m)* ····→

EIFFEL TOWER
←··· *1,063 ft (324 m)* ···→

OVERBURDEN CONVEYOR BRIDGE F60
←················ *1,647 ft (502 m)* ················→

IS IT?

GIRAFFE
19 ft (5.8 m)

FIRST-GRADER
3.6 ft (1.1 m)

SEMITRAILER TRUCK
◄ ······ *70 ft (21.3 m)* ······ ►

TITANIC
◄ ······ *883 ft (269 m)* ······ ►

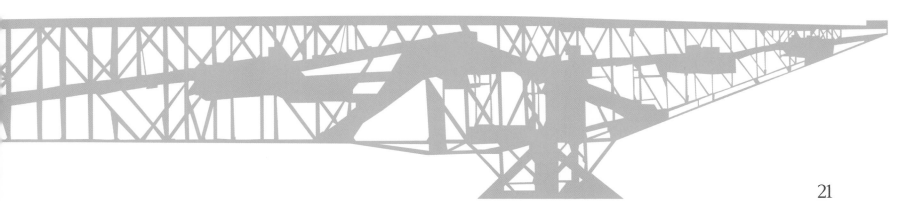

GLOSSARY

conveyor belts—*moving bands that carry materials from one place to another*

open-pit mines—*mines where machines dig into the ground from the surface*

overburden—*the soil, rocks, and other materials above an underground store of minerals, such as coal*

READ MORE

Allen, Kenny. *Earthmovers.*
New York: Gareth Stevens, 2013.

Zappa, Marcia. *Fossil Fuels.*
Edina, Minn.: Abdo, 2011.

WEBSITES

Energy Kids: Nonrenewable Coal
http://www.eia.gov/kids/energy.cfm?page=coal_home-basics
Learn all about how coal is formed, the different types of coal, and coal mining.

Energy Star Kids
http://www.energystar.gov/index.cfm?c=kids.kids_index
Find out where energy comes from and how you can save energy each day.

Note: Every effort has been made to ensure that the websites listed above are suitable for children, that they have educational value, and that they contain no inappropriate material. However, because of the nature of the Internet, it is impossible to guarantee that these sites will remain active indefinitely or that their contents will not be altered.

INDEX